Personal Math
Companion

Personal Math Companion

Marsha P. Smith

Illustrated by Brittani Brown

To order additional copies of this book, contact:
Xlibris Corporation
1-888-795-4274
www.Xlibris.com
Orders@Xlibris.com
80911

Contents

General Mathematical Concepts

Algebra

Geometry

Statistics & Probability

Glossary

NUMBERS

NATURAL NUMBERS:
1, 2, 3, 4, 5, 6, 7, 8, 9, etc.

EVEN NUMBERS:
2, 4, 6, 8, 10, 12, 14, 16, 18, etc.

ODD NUMBERS:
1, 3, 5, 7, 9, 11, 13, 15, 17, etc.

PRIME NUMBERS:
2, 3, 5, 7, 11, 13, 17, 19, 23, etc.

COMPOSITE NUMBERS:
4, 6, 8, 9, 10, 12, 14, 15, 16, etc.

PLACE VALUES

1 0 0 0 0 0 0 0 0 0

Ones	(O)
Tens	(T)
Hundreds	(H)
Thousands	(TH)
Ten Thousands	(TTH)
Hundred Thousands	(HTH)
Millions	(M)
Ten Millions	(TM)
Hundred Millions	(HM)
Billions	(B)

EXPANDED FORM

Example: 1, 987,654,321

B	HM	TM	M	HTH	TTH	TH	H	T	O
1	0	0	0	0	0	0	0	0	0
	9	0	0	0	0	0	0	0	0
		8	0	0	0	0	0	0	0
			7	0	0	0	0	0	0
				6	0	0	0	0	0
					5	0	0	0	0
						4	0	0	0
							3	0	0
								2	0
+									1
1,	9	8	7,	6	5	4,	3	2	1

POWERS OF TEN

10^1	10	Ten
10^2	100	Hundred
10^3	1,000	Thousand
10^4	10,000	Ten Thousand
10^5	100,000	Hundred Thousand
10^6	1,000,000	Million
10^7	10,000,000	Ten Million
10^8	100,000,000	Hundred Million
10^9	1,000,000,000	Billion

SMALLEST & LARGEST NATURAL NUMBERS

Number of Digits	Smallest Number	Largest Number
1	1	9
2	10	99
3	100	999
4	1,000	9,999
5	10,000	99,999
6	100,000	999,999
7	1,000,000	9,999,999
8	10,000,000	99,999,999
9	100,000,000	999,999,999

ROMAN NUMERAL		
	Hindu Arabic Numerals	Roman Numerals
One	1	I
Two	2	II
Three	3	III
Four	4	IV
Five	5	V
Six	6	VI
Seven	7	VII
Eight	8	VIII
Nine	9	IX
Ten	10	X
Eleven	11	XI
Twelve	12	XII
Thirteen	13	XIII
Fourteen	14	XIV
Fifteen	15	XV

ROMAN NUMERAL continued		
	Hindu Arabic Numerals	Roman Numerals
Sixteen	16	XVI
Seventeen	17	XVII
Eighteen	18	XVIII
Nineteen	19	XIX
Twenty	20	XX
Thirty	30	XXX
Forty	40	XL
Fifty	50	L
Sixty	60	LX
Seventy	70	LXX
Eighty	80	LXXX
Ninety	90	XC
Hundred	100	C
Four Hundred	400	CD
Five Hundred	500	D
Nine Hundred	900	CM
Thousand	1,000	M

2,3,4,5 TIME TABLE

2 TIMES	3 TIMES	4 TIMES	5 TIMES
2 x 1 = 2	3 x 1 = 3	4 x 1 = 4	5 x 1 = 5
2 x 2 = 4	3 x 2 = 6	4 x 2 = 8	5 x 2 = 10
2 x 3 = 6	3 x 3 = 9	4 x 3 = 12	5 x 3 = 15
2 x 4 = 8	3 x 4 = 12	4 x 4 = 16	5 x 4 = 20
2 x 5 = 10	3 x 5 = 15	4 x 5 = 20	5 x 5 = 25
2 x 6 = 12	3 x 6 = 18	4 x 6 = 24	5 x 6 = 30
2 x 7 = 14	3 x 7 = 21	4 x 7 = 28	5 x 7 = 35
2 x 8 = 16	3 x 8 = 24	4 x 8 = 32	5 x 8 = 40
2 x 9 = 18	3 x 9 = 27	4 x 9 = 36	5 x 9 = 45
2 x 10 = 20	3 x 10 = 30	4 x 10 = 40	5 x 10 = 50
2 x 11 = 22	3 x 11 = 33	4 x 11 = 44	5 x 11 = 55
2 x 12 = 24	3 x 12 = 36	4 x 12 = 48	5 x 12 = 60

6,7,8,9 TIME TABLE

6 TIMES	7 TIMES	8 TIMES	9 TIMES
6 x 1 = 6	7 x 1 = 7	8 x 1 = 8	9 x 1 = 9
6 x 2 = 12	7 x 2 = 14	8 x 2 = 16	9 x 2 = 18
6 x 3 = 18	7 x 3 = 21	8 x 3 = 24	9 x 3 = 27
6 x 4 = 24	7 x 4 = 28	8 x 4 = 32	9 x 4 = 36
6 x 5 = 30	7 x 5 = 35	8 x 5 = 40	9 x 5 = 45
6 x 6 = 36	7 x 6 = 42	8 x 6 = 48	9 x 6 = 54
6 x 7 = 42	7 x 7 = 49	8 x 7 = 56	9 x 7 = 63
6 x 8 = 48	7 x 8 = 56	8 x 8 = 64	9 x 8 = 72
6 x 9 = 54	7 x 9 = 63	8 x 9 = 72	9 x 9 = 81
6 x 10 = 60	7 x 10 = 70	8 x 10 = 80	9 x 10 = 90
6 x 11 = 66	7 x 11 = 77	8 x 11 = 88	9 x 11 = 99
6 x 12 = 72	7 x 12 = 84	8 x 12 = 96	9 x 12 = 108

10,11,12 TIMES TABLE		
10 TIMES	**11 TIMES**	**12 TIMES**
10 x 1 = 10	11 x 1 = 11	12 x 1 = 12
10 x 2 = 20	11 x 2 = 22	12 x 2 = 24
10 x 3 = 30	11 x 3 = 33	12 x 3 = 36
10 x 4 = 40	11 x 4 = 44	12 x 4 = 48
10 x 5 = 50	11 x 5 = 55	12 x 5 = 60
10 x 6 = 60	11 x 6 = 66	12 x 6 = 72
10 x 7 = 70	11 x 7 = 77	12 x 7 = 84
10 x 8 = 80	11 x 8 = 88	12 x 8 = 96
10 x 9 = 90	11 x 9 = 99	12 x 9 = 108
10 x 10 = 100	11 x 10 = 110	12 x 10 = 120
10 x 11 = 110	11 x 11 = 121	12 x 11 = 132
10 x 12 = 120	11 x 12 = 132	12 x 12 = 144

SQUARE NUMBERS AND ROOTS

A **square number** is the result of a number that has been multiplied by itself; so $6^2 = (6 \times 6)$; thus the squared number $= \sqrt{36} = 6$ $7^2 = (7 \times 7)$; thus $\sqrt{49} = 7$
We say the square of 6 is 36; thus the "**square root**" (radical) of 36 is 6

Radical symbol $\sqrt{36} = 6$ Root of 36

Multiplying Radicals

Rule for *multiplying* radicals $\sqrt{a} \times \sqrt{b} = \sqrt{ab}$
Examples:

(a) $\sqrt[3]{6} \times \sqrt[4]{7} = \sqrt[12]{42}$ (b) $\sqrt{2} \times \sqrt{18} = \sqrt{36} = 6$

Dividing Radicals

Rule for *dividing* radicals $\quad\quad \dfrac{\sqrt{a}}{\sqrt{b}} = \sqrt{\dfrac{a}{b}}$

Examples: (a) $\quad \dfrac{\sqrt{8}}{\sqrt{4}} = \sqrt{2}$ $\quad\quad\quad$ (b) $\dfrac{\sqrt[3]{12}}{\sqrt[6]{3}} = \dfrac{1}{2}\sqrt{4}$

SQUARE ROOT TABLE					
1^2	=	1	16^2	=	256
2^2	=	4	17^2	=	289
3^2	=	9	18^2	=	324
4^2	=	16	19^2	=	361
5^2	=	25	20^2	=	400
6^2	=	36	21^2	=	441
7^2	=	49	22^2	=	484
8	=	64	23^2	=	529
9^2	=	81	24^2	=	576
10^2	=	100	25^2	=	625
11^2	=	121	26^2	=	676
12^2	=	144	27^2	=	729
13^2	=	169	28^2	=	784
14^2	=	196	29^2	=	841
15^2	=	225	30^2	=	900

FRACTIONS

A fraction is a number that represents part of a whole

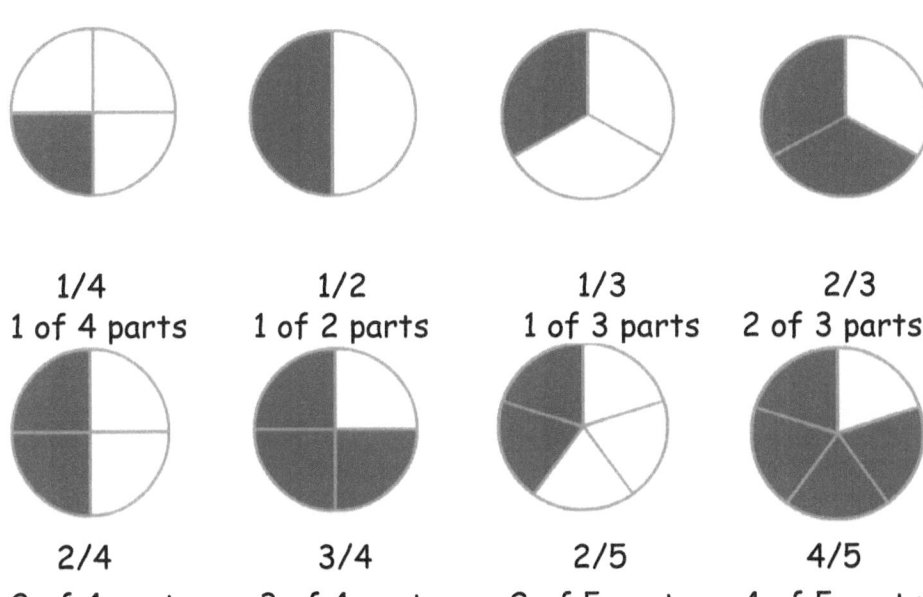

1/4	1/2	1/3	2/3
1 of 4 parts	1 of 2 parts	1 of 3 parts	2 of 3 parts

2/4	3/4	2/5	4/5
2 of 4 parts	3 of 4 parts	2 of 5 parts	4 of 5 parts

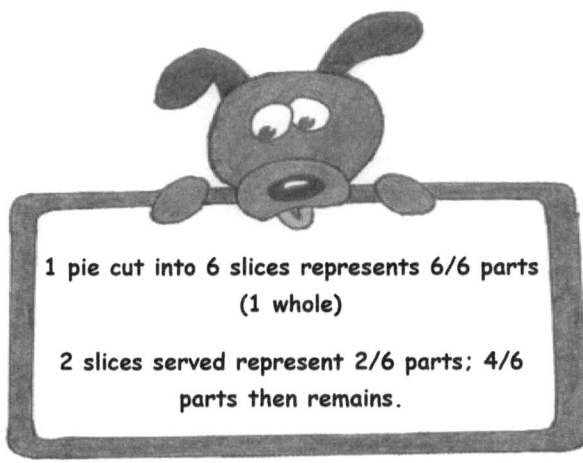

1 pie cut into 6 slices represents 6/6 parts
(1 whole)

2 slices served represent 2/6 parts; 4/6
parts then remains.

FRACTIONS continued

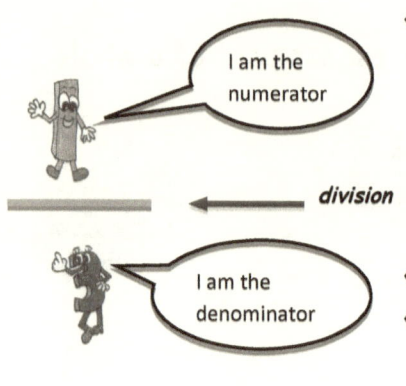

I am the numerator

division

I am the denominator

- ***Reciprocal*** of is fraction is the fraction turn upside down

  eg. The reciprocal of $\underline{3}$ is $\underline{4}$ *this is only*
  $\qquad\qquad\qquad\quad 4 \qquad 3$

  done in division of fractions

- **A decimal** is a fraction

- Any fraction with zero (0) as the denominator is undefined

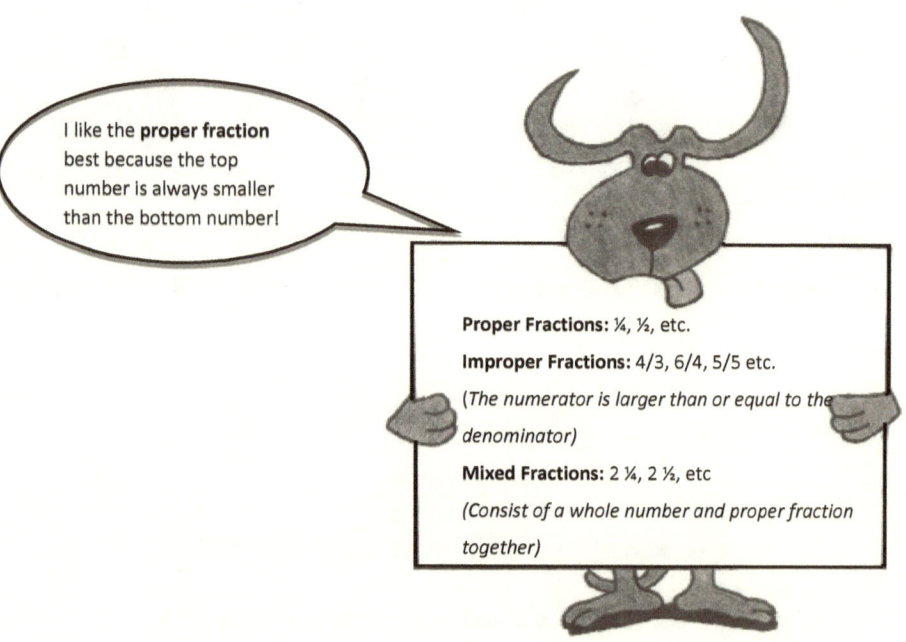

I like the **proper fraction** best because the top number is always smaller than the bottom number!

Proper Fractions: ¼, ½, etc.

Improper Fractions: 4/3, 6/4, 5/5 etc.

(The numerator is larger than or equal to the denominator)

Mixed Fractions: 2 ¼, 2 ½, etc

(Consist of a whole number and proper fraction together)

- *To **multiply fractions**, multiply the numerator of one fraction by the numerator of the other, repeat the same step for the denominators*

Example
$$\frac{1}{2} \times \frac{3}{4} = \frac{3}{8}$$

- *To **divide** a whole number or a fraction, multiply the whole number or fraction by the <u>reciprocal</u> of the fraction.*

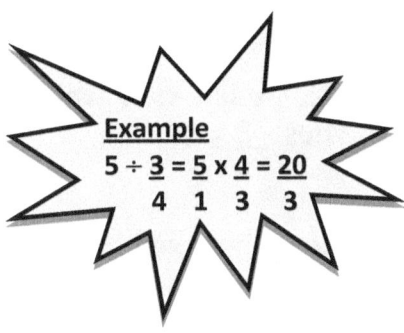

<u>Example</u>
$$5 \div \frac{3}{4} = \frac{5}{1} \times \frac{4}{3} = \frac{20}{3}$$

- To *convert a **fraction** to a **decimal***, divide the numerator by the denominator.

<u>Example</u>
$$\frac{3}{4} = 0.75 \text{ or } \frac{1}{2} = 0.5$$

- To *convert a **fraction** to an **equivalent percentage**,* multiply the fraction by 100

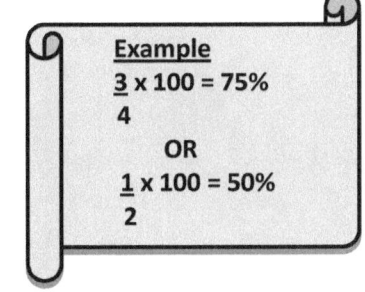

<u>Example</u>
$$\frac{3}{4} \times 100 = 75\%$$

OR

$$\frac{1}{2} \times 100 = 50\%$$

- To *convert a **percentage** to an **equivalent fraction**,* divide the percentage by 100

Example
75% ÷ 100 = 0.75 or 50% ÷ 100 = 0.5

- *In order to convert a **mixed fraction** into an **improper fraction**, multiply the whole number by the denominator then add the numerator.*

RATIO, PROPORTION & RATE

Ratio: a comparison of two quantities with the same units. Three different ways to write ratios:

3:4   3 to 4   3/4

Note
A ratio must always have 2 numbers and be in its simplest form.

Proportion: is an equation with a ratio on each side which states that the two ratios are equal, thus: $\dfrac{2}{4} = \dfrac{4}{8}$

Rate: is a ratio that expresses how long it takes to get something done, such as traveling from home to school.

Eg. What amount of time will it take a car to travel 84 miles at an average speed of 48 miles per hour?

$$\text{Speed} = \dfrac{84 \text{ miles}}{48 \text{ miles}} = 1\dfrac{3}{4} \text{ h}$$

The Average Speed (rate) = the distance traveled
The time taken

DECIMAL SYSTEM

In the decimal system we count in base 10 and use the ten digits 0 to 9. Since each number base or scale is 10, each digit of a number has a place value in terms of powers of 10.

Thus:
The number 983275_{10} can be represented as follow:

Place Name	Hundred thousands	Ten thousands	Thousands	Hundreds	Tens	Ones
Place value	100 000 $=10^5$	10 000 $=10^4$	1000 $=10^3$	100 $=10^2$	10 $=10^1$	1 $= 10^0$
Digit	9	8	3	2	7	5

$$= 9 \times 10^5 + 8 \times 10^4 + 3 \times 10^3 + 2 \times 10^2 + 7 \times 10^1 + 5 \times 10^0$$

The number 0.4601_{10} can be represented as follow:

Place name	Tenths	Hundredths	Thousandths	Ten thousandths
Place value	$\frac{1}{10} = 0.1$ $= 10^{-1}$	$\frac{1}{100} = 0.01$ $= 10^{-2}$	$\frac{1}{0.001\ 1000} = $ $= 10^{-3}$	$\frac{1}{10\ 000} = 0.0001$ $= 10^{-4}$
Digit	4	6	0	1

$$= 4 \times 10^{-1} + 6 \times 10^{-2} + 0 \times 10^{-3} + 1 \times 10^{-4}$$
$$= 4 \times 10^{-1} + 6 \times 10^{-2} + 1 \times 10^{-4}$$

Hence the number 983275.4601_{10} can be represented as:
$9 \times 10^5 + 8 \times 10^4 + 3 \times 10^3 + 2\ 10^2 + 7 \times 10^1 + 5 \times 10^0 + 4 \times 10^{-1} + 6 \times 10^{-2} + 0 \times 10^{-3} + 1 \times 10^{-4}$

> **Please Note:** For any number, powers keep increasing by increments of one moving to the left, from digit to digit: and decreasing by increments of one, moving to the right, from digit to digit. The decimal point separates the positive indices and zero index from the negative indices of the place values

COMPUTING DECIMALS

Adding & Subtracting Decimals

When adding or subtracting decimals, line up the decimal points and follow the rules for adding and subtracting whole numbers and place the decimal point in the same place as above.

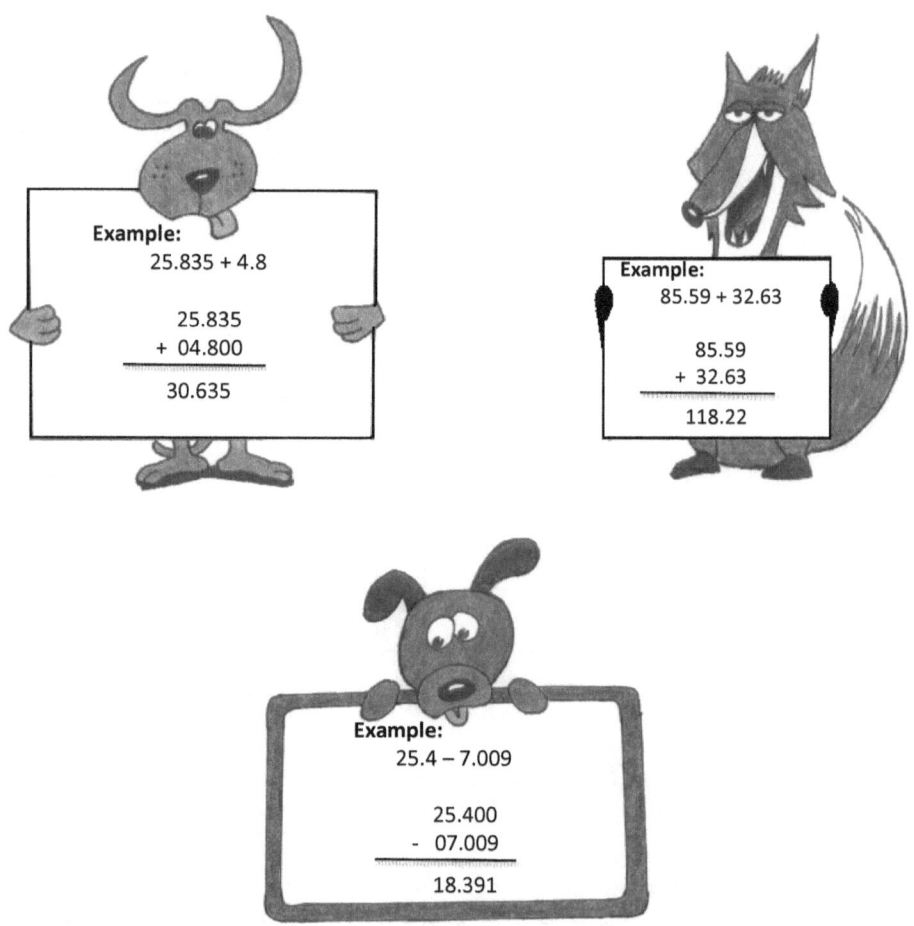

Example:
25.835 + 4.8

```
  25.835
+ 04.800
_____
  30.635
```

Example:
85.59 + 32.63

```
   85.59
 + 32.63
_____
  118.22
```

Example:
25.4 – 7.009

```
  25.400
- 07.009
_____
  18.391
```

Remember: *If one number has more decimal places than the others, use zero's (0) to give them the same number of decimal places.*

Multiplying Decimals

When we multiply decimals, we use the same rules as multiplying whole numbers. The only difference is we need to decide how many digits should be on the right side of the decimal place.

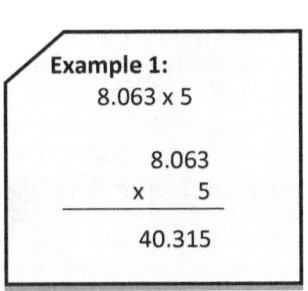

Example 1:
8.063 x 5

8.063
x   5
─────────
40.315

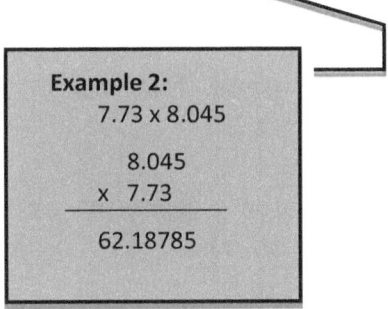

Example 2:
7.73 x 8.045

8.045
x 7.73
─────────
62.18785

Note: *To determine how many digits to leave to the right of the decimal point, we count the number of digits to the right of the decimal point in both factors.*

In **example 1** there are three (3) digits to the right of the decimal point.

In **example 2** there are five (5) digits total to the right of both numbers, (3 at the top and 2 at the bottom), which means, in the answers we put 5 numbers to the right of the decimal point.

When we **write a decimal as a percent**, we move the decimal point two places to the right, example 0.08 when the point is moved 2 places to the right becomes 8%

Dividing Decimals by a whole number

When we divide decimals by a whole number, we use the same rules as long division. Keep the decimal point lined up.

Example: $14\overline{)16.63}$ with answer 1.18 remainder **11**

Note: If asked to round your answer to a specific place, divide one place further than the place you are rounding to. If necessary, add zeros (0) to the number being divided.

Dividing Decimals by Decimals

When dividing a decimal, multiply the decimal by a power of 10 that is large enough to create a whole number. Using the same power of 10, repeat the same step on the dividend, the problem becomes division of whole numbers.

Example: 0.288 ÷ 0.24 **(when we multiply by 100, for each (0), move the decimal point to the right)**; Thus 28.8 ÷ 24 = 1.2

When we **write a percent as a decimal** we move the decimal two places to the left, example 42% becomes .42; 1.87% becomes 0.0187

Lowest Common Multiplies (LCM)

LCM is the smallest number that the given numbers will divide into without a remainder (*a multiple of two or more numbers*).

Example: The LCM of 3 and 5 is **15**;

seeing 3 x 5 = 15 and 5 x 3 = 15, it means that 15 is a multiple of 3 and 5. Other common multiples include 30 and 45, seeing 3 and 5 will divide into these numbers but they are not the lowest multiple we say that the **LCM** of 3 and 5 is 15

Greatest Common Factors (GCF)

GCF of two or more whole numbers is the **largest whole number** that will divide evenly into each of the numbers. To find the GCF we list all the factors of each number; list the **common factors** then choose the largest one.

Example: The *GCF* of 36 and 54; we list all the factors of 36 and 54
36 {1, 2, 3, 4, 6, 9, 12, 18, and 36}   54 {1, 2, 3, 6, 9, 18, 27, and 54.}

The **common factors** are therefore 1, 2, 3, 6, 9, 18 with **18** being the highest factor of the two numbers.

Prime Factorization:
When determining prime factorization, we need to find the **prime factors** in the number. Once this is done, when we multiply these prime numbers, the product should be the result of the original number.

Example: If asked to find the prime factors of 12. We start by dividing the number by the smallest prime number and work the number until the total is "1"

$$12 \div 2 = 6$$
$$6 \div 2 = 3$$
$$3 \div 3 = 1$$

The factors of 12 are 2 x 2 x 3 therefore 12 = 2 x 2 x 3

Note all the numbers used as divisors are prime numbers

Example: find the prime factorization of 147
Seeing we cannot divide 147 evenly by 2, we will try the next number up "3"
We see that 147 $\div$ 3 = 49
We then factor 49 (the smallest number that will divide into 49 is 7)
Therefore to find the prime factors of 147 see below:

$$147 \div 3 = 49$$
$$49 \div 7 = 7$$
$$7 \div 7 = 1$$

Thus the factors and prime numbers of 147 = 3 x 7 x 7

ROUNDING AND ESTIMATING NUMBERS

When rounding off numbers, the digit to the right of the number being rounded, determines if the number is rounded up or down.

If the number to the right is 1, 2, 3, or 4 then *round down*.

If the number is 5, 6, 7, 8 or 9 then *round up*.

Round Down   **Round Up**

1	2	3	4	5	6	7	8	9	10
11	12	13	14	15	16	17	18	19	20
21	22	23	24	25	26	27	28	29	30
31	32	33	34	35	36	37	38	39	40
41	42	43	44	45	46	47	48	49	50
51	52	53	54	55	56	57	58	59	60
61	62	63	64	65	66	67	68	69	70
71	72	73	74	75	76	77	78	79	80
81	82	83	84	85	86	87	88	89	90
91	92	93	94	95	96	97	98	99	100

eg.   3,825 *rounded to 3 **significant figures** is* 3,830
2,840 *rounded to the **nearest hundred** is* 2,800
9.374 *rounded to 3 **decimal places** is* 9.37
9.420 *rounded to 1 decimal place is* 9.4
7.8 *rounded to the **nearest whole number** is* 8

Rational Numbers

Rational numbers: numbers that can be written as a simple fraction (ratio).

Rule: p/q where **p** and **q** are integers and **q** is not equal to zero (0)

Example: 3.5 can be written as a fraction 7/2

Sample numbers and their fraction

Number	Fraction
7	7/1
2.5	9/2
.5	1/2
0.33...	1/3

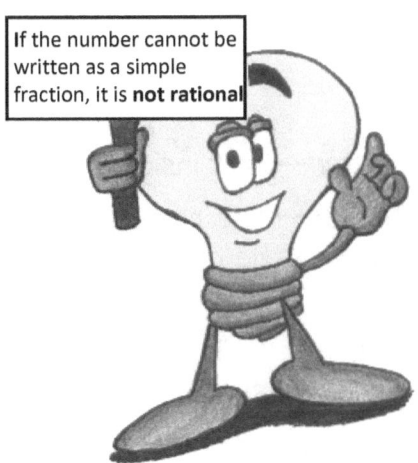

If the number cannot be written as a simple fraction, it is **not rational**

Irrational Numbers

Irrational numbers: numbers that **cannot** be written as a simple fraction or ratio

Common irrational number

$$\sqrt{2} \qquad \sqrt{3} \qquad \sqrt{5} \qquad \pi\text{(pi)}$$

Note the above numbers cannot be written as a simple fraction/ratio (p/q), however $\sqrt{4}$ can, seeing $\sqrt{4}$ = 2 and 2 can be written as 2/1 so it is rational. The others have no root or go on indefinitely.

PROPERTIES OF A CIRCLE

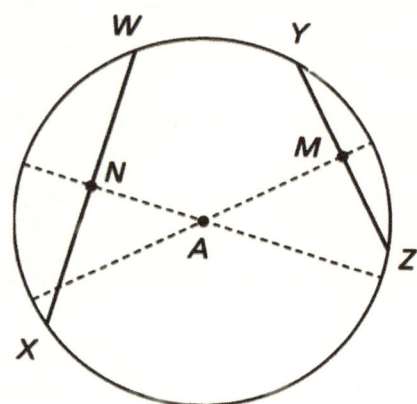

Center—point inside the circle (in this diagram it is labeled "A")

Radius—The radius is the distance from the center to any point on the circle. It is half the diameter (*note the broken lines*)

Diameter—The distance across the circle, the length of any chord passing through the center and is twice the radius.

Circumference—The circumference is the distance around the circle

Area—Note that a circle is a closed line, which means it does not have an area. However, the space within the enclosed line (inside the circle) is referred to as the area.

Chord—A line segment linking any two points on a circle

Pi—The circumference of any circle, divided by its diameter (3.142)

See page 60 for finding the following
Area of a circle
Circumference
Area of a semicircle
Diameter

SOME EQUIVALENTS

Fractions	Decimals	Percentages
1/100	0.01	1%
1/50	0.02	2%
1/40	0.025	2.5%
1/25	0.04	4%
1/20	0.05	55%
1/10	0.10	10%
1/8	0.125	12.5%
3/8	0.375	37.5%
5/8	0.625	62.5%
7/8	0.875	87.5%
1/5	0.20	20%
2/5	0.40	40%
3/5	0.60	60%
4/5	0.80	80%
1/4	0.25	25%
3/4	0.75	75%
1/3	0.33	33.3%
2/3	0.66	66.6%
1/2	0.50	50%

MARSHA P. SMITH

STANDARD FRACTIONS MULTIPLICATION TABLE

$\frac{1}{8}$	$\frac{1}{2}$	$\frac{3}{4}$	$\frac{1}{4}$
$1 \times \frac{1}{8} = 11/8$ $10 \times \frac{1}{8} = 1\frac{1}{4}$ $100 \times \frac{1}{8} = 12\frac{1}{2}$	$1 \times \frac{1}{2} = \frac{1}{2}$ $10 \times \frac{1}{2} = 5$ $100 \times \frac{1}{2} = 50$	$1 \times \frac{3}{4} = \frac{3}{4}$ $10 \times \frac{3}{4} = 7\frac{1}{2}$ $100 \times \frac{3}{4} = 75$	$1 \times \frac{1}{4} = \frac{1}{4}$ $10 \times \frac{1}{4} = 2\frac{1}{2}$ $100 \times \frac{1}{4} = 25$
$2 \times \frac{1}{8} = \frac{1}{4}$ $20 \times \frac{1}{8} = 2\frac{1}{2}$ $200 \times \frac{1}{8} = 25$	$2 \times \frac{1}{2} = 1$ $20 \times \frac{1}{2} = 10$ $200 \times \frac{1}{2} = 100$	$2 \times \frac{3}{4} = 1\frac{1}{2}$ $20 \times \frac{3}{4} = 15$ $200 \times \frac{3}{4} = 150$	$2 \times \frac{1}{4} = \frac{1}{2}$ $20 \times \frac{1}{4} = 5$ $200 \times \frac{1}{4} = 50$
$3 \times \frac{1}{8} = 3/8$ $30 \times \frac{1}{8} = 33/4$ $300 \times \frac{1}{8} = 37\frac{1}{2}$	$3 \times \frac{1}{2} = 1\frac{1}{2}$ $30 \times \frac{1}{2} = 15$ $300 \times \frac{1}{2} = 150$	$3 \times \frac{3}{4} = 2\frac{1}{4}$ $30 \times \frac{3}{4} = 22\frac{1}{2}$ $300 \times \frac{3}{4} = 225$	$3 \times \frac{1}{4} = \frac{3}{4}$ $30 \times \frac{1}{4} = 7\frac{1}{2}$ $300 \times \frac{1}{4} = 75$
$4 \times \frac{1}{8} = \frac{1}{2}$ $40 \times \frac{1}{8} = 5$ $400 \times \frac{1}{8} = 50$	$4 \times \frac{1}{2} = 2$ $40 \times \frac{1}{2} = 20$ $400 \times \frac{1}{2} = 200$	$4 \times \frac{3}{4} = 3$ $40 \times \frac{3}{4} = 30$ $400 \times \frac{3}{4} = 300$	$4 \times \frac{1}{4} = 1$ $40 \times \frac{1}{4} = 10$ $400 \times \frac{1}{4} = 100$
$5 \times \frac{1}{8} = 5/8$ $50 \times \frac{1}{8} = 6\frac{1}{4}$ $500 \times \frac{1}{8} = 61\frac{1}{2}$	$5 \times \frac{1}{2} = 2\frac{1}{2}$ $50 \times \frac{1}{2} = 25$ $500 \times \frac{1}{2} = 250$	$5 \times \frac{3}{4} = 3\frac{3}{4}$ $50 \times \frac{3}{4} = 37\frac{1}{2}$ $500 \times \frac{3}{4} = 375$	$5 \times \frac{1}{4} = 1\frac{1}{4}$ $50 \times \frac{1}{4} = 12\frac{1}{2}$ $500 \times \frac{1}{4} = 125$
$6 \times \frac{1}{8} = 3/4$ $60 \times \frac{1}{8} = 7\frac{1}{2}$ $600 \times \frac{1}{8} = 75$	$6 \times \frac{1}{2} = 3$ $60 \times \frac{1}{2} = 30$ $600 \times \frac{1}{2} = 300$	$6 \times \frac{3}{4} = 4\frac{1}{2}$ $60 \times \frac{3}{4} = 45$ $600 \times \frac{3}{4} = 450$	$6 \times \frac{1}{4} = 1\frac{1}{2}$ $60 \times \frac{1}{4} = 15$ $600 \times \frac{1}{4} = 150$
$7 \times \frac{1}{4} = 1\frac{3}{4}$ $70 \times \frac{1}{4} = 17\frac{1}{2}$ $700 \times \frac{1}{4} = 175$	$7 \times \frac{1}{2} = 3 1/2$ $70 \times \frac{1}{2} = 35$ $700 \times \frac{1}{2} = 350$	$7 \times \frac{3}{4} = 5\frac{1}{4}$ $70 \times \frac{3}{4} = 52\frac{1}{2}$ $700 \times \frac{3}{4} = 525$	$7 \times \frac{1}{8} = 7/8$ $70 \times \frac{1}{8} = 8\frac{3}{4}$ $700 \times \frac{1}{8} = 87\frac{1}{2}$
$8 \times \frac{1}{4} = 2$ $80 \times \frac{1}{4} = 20$ $800 \times \frac{1}{4} = 200$	$8 \times \frac{1}{2} = 4$ $80 \times \frac{1}{2} = 40$ $800 \times \frac{1}{2} = 400$	$8 \times \frac{3}{4} = 6$ $80 \times \frac{3}{4} = 60$ $800 \times \frac{3}{4} = 600$	$8 \times \frac{1}{8} = 1$ $80 \times \frac{1}{8} = 10$ $800 \times \frac{1}{8} = 100$
$9 \times \frac{1}{4} = 2\frac{1}{4}$ $90 \times \frac{1}{4} = 22\frac{1}{2}$ $900 \times \frac{1}{4} = 225$ $1000 \times \frac{1}{4} = 250$	$9 \times \frac{1}{2} = 4\frac{1}{2}$ $90 \times \frac{1}{2} = 45$ $900 \times \frac{1}{2} = 450$ $1000 \times \frac{1}{2} = 500$	$9 \times \frac{3}{4} = 6\frac{3}{4}$ $90 \times \frac{3}{4} = 67\frac{1}{2}$ $900 \times \frac{3}{4} = 675$ $1000 \times \frac{3}{4} = 750$	$9 \times \frac{1}{8} = 1\frac{1}{8}$ $90 \times \frac{1}{8} = 11\frac{1}{4}$ $900 \times \frac{1}{8} = 112\frac{1}{2}$ $1000 \times \frac{1}{8} = 125$

CONSUMER MATH

Compounding Averages: To find the average of a set of quantities, add them together and divide the total by the number of quantities.

Profit & Loss

a. When an article is bought at a certain price and sold at a higher price, the amount gained is called the **profit**.

Profit   = Selling Price - Cost Price

Profit Percent = $\dfrac{\text{Selling Price - Cost Price}}{\text{Cost Price}} \times 100$

b. When an article is bought at a certain price and sold at a lower price, the amount lost is called the **loss**.

Loss   = Cost Price - Selling Price

Loss percent = $\dfrac{\text{Cost Price - Selling Price}}{\text{Cost Price}} \times 100$

c. **Selling Price**   =   Cost Price + Profit

Finding Average

To find the **average** add all the numbers together, then divide them by the amount of numbers there are.

$$3 + 6 + 7 + 10 + 12 + 15 + 17 = 70$$

The total numbers added are 7

Next you take the sum (70) and divide by the amount of numbers (7)

$$\text{Average} = 70 \div 7 = 10$$

MEASURES OF WEIGHT

Imperial System

16 Ounces	=	1 Pound
14 Pounds	=	1 Stone
28 Pounds	=	1 Quarter
4 Quarters	=	1 Hundredweight
20 Hundredweights	=	1 Ton
112 Pounds	=	1 Hundredweight
2240 Pounds	=	1 Ton

Conversion

28.35 Grams	=	1 Ounce
0.45 Kilogram	=	1 Pound
2.20 Pounds	=	1 Kilogram
0.98 Ton	=	1 Metric Ton

Metric System

10 Milligrams	=	1 Centigram
10 Centigrams	=	1 Decigram
10 Decigrams	=	1 Gram
10 Grams	=	1 Decagram
10 Decagram	=	1 Hectogram
10 Hectograms	=	1 Kilogram
1000 Grams	=	1 Kilogram
1000 Kilograms	=	1 Metric Ton

MEASURES OF LENGTH

Imperial System

12 Inches	=	1 Foot
3 Feet	=	1 Yard
22 yards	=	1 Chain
10 Chains	=	1 Furlong
8 Furlongs	=	1 Mile
220 Yards	=	1 Furlong
1760 Yards	=	1 Mile
5280 Feet	=	1 Mile

Conversions

2.5 Centimeters	=	1 Inch
30 Centimeters	=	1 Foot
0.9 Meters	=	1 Yard
1.6 Kilometers	=	1 Mile
0.04 Inches	=	1 Millimeter
0.4 Inches	=	1 Centimeter
3.3 Feet	=	1 Meter
1.1 Yards	=	1 Meter
0.6 Miles	=	1 Kilometer
201.17 Meters	=	1 Furlong

Metric System

10 Millimeters	=	1 Centimeter
10 Centimeter	=	1 Decimeter
10 Decimeters	=	1 Meter
10 Meters	=	1 Decameter
10 Decameters	=	1 Hectometer
10 Hectometers	=	1 Kilometer
100 Centimeters	=	1 Meter
1000 Meters	=	1 Kilometer

MEASURES OF ACRE

Imperial System

144 Square Inches	=	1 Square Foot
9 Square Feet	=	1 Square Yard
4840 Square Yards	=	1 Acre
640 Acres	=	1 Square Mile

Conversions

6.5 Square Centimeters	=	1 Square Inch
0.09 Square Meters	=	1 Yard
0.8 Square Meters	=	1 Square Yard
2.6 Square Kilometers	=	1 Square Mile
0.4 Hectares	=	1 Acre
0.16 Square Inches	=	1 Square Centimeter
1.2 Square Yards	=	1 Square Meter
10.76 Square Feet	=	1 Square Meter
0.4 Square Miles	=	1 Square Kilometer
2.24 Acres	=	1 Hectare (10,000 square meters)

Metric System

100 Square Millimeters	=	1 Square Centimeter
10,000 Square Centimeters	=	1 Square Meter
100 Square Meters	=	1 Acre
100 Acres	=	1 Hectare
10,000 Square Meters	=	1 Hectare
100 Hectares	=	1 Square Kilometer

MEASURES OF MASS

28 Grams	=	1 Ounce
0.45 Kilograms	=	1 Pound
0.9 Metric ton	=	1 Short ton or 2000 pounds
0.035 Ounces	=	1 Gram
2.2 Pounds	=	1 Kilogram
1.1 Short ton	=	1 Metric ton (1,000 kilograms)

MEASURES OF DEPTH

3.3 Meters	=	1 Foot
6 Feet	=	1 Fathom
120 Fathoms	=	1 Cable

SURVEY

7.92 Inches	=	1 Link
16.5 Feet	=	1 Rod
4 Rod	=	1 Chain
10 Chain	=	1 Furlong
8 Furlong	=	1 Statute mile
3 Miles	=	1 League

MEASURES OF VOLUME

5 Milliliters	=	1 Teaspoon
15 Milliliters	=	1 Tablespoon
16 Milliliters	=	1 Cubic Inch
30 Milliliters	=	1 Fluid Ounce
0.24 Liters	=	1 Cup
0.47 Liters	=	1 Pint
0.95 Liters	=	1 Quart
3.8 Liters	=	1 Gallon
0.03 Cubic Meters	=	1 Cubic Feet
0.76 Cubic Meters	=	1 Cubic Yards
0.03 Fluid Ounces	=	1 Milliliter
0.06 Cubic Inches	=	1 Milliliter
2.1 Pints	=	1 Liter
1.06 Quarts	=	1 Liter
0.26 Gallons	=	1 Liter
35 Cubic Feet	=	1 Cubic Meter
1.3 Cubic Yards	=	1 Cubic Meter

DRY VOLUME

33.6 Cubic inches	=	1 Pint
2 Pints	=	1 Quart
4 Quarts	=	1 Gallon
2 Gallons	=	1 Peck
3.281 Bushel	=	1 Barrel
4 Pecks	=	1 Bushel

MEASURES OF PAPER

24 Sheets	=	1 Quire
20 Quires	=	1 Ream
10 Reams	=	1 Bale

MEASURES OF CAPACITY *liquid*

Imperial System

8 Ounces	=	1 Cup
2 Cups	=	1 Pint
2 Pints	=	1 Quart
128 Ounces	=	1 Gallon
4 Quarts	=	1 Gallon
4 Gills	=	1 Pint
2 Pints	=	1 Quart
4 Quarts	=	1 Gallon
$31\frac{1}{2}$ Gallons	=	1 Barrel

Metric System

10 Milliliters	=	1 Centiliter
10 Centiliters	=	1 Deciliter
10 Deciliters	=	1 Liter
1000 Milliliters	=	1 Liter
1000 Liters	=	1 Kiloliter

CUSTOMARY LIQUID & DRY MEASUREMENTS

1.25 Milliliters	=	1/4 Teaspoon
2.5 Milliliters	=	1/2 Teaspoon
5 Milliliters	=	1 Teaspoon
15 Milliliters	=	1 Tablespoon
30 Milliliters	=	1 Fluid ounce
60 Milliliters	=	1/4 Cup
80 Milliliters	=	1/3 Cup
120 Milliliters	=	1/2 Cup
240 Milliliters	=	1 Cup
480 Milliliters	=	1 Pint
960 Milliliters	=	1 Quart
3.84 Liters	=	1 Gallon
28 Grams	=	1 Ounce
114 Grams	=	1/4 Pound
454 Grams	=	1 Pound
1 Kilogram	=	2.2 Pounds

GRAIN MEASUREMENT

64.79891 mg	=	1 Grain
1.771845 g	=	1 Dram
28.34952 g	=	1 Ounce
453.59237 g	=	1 Pound
45.359237 kg	=	1 US Hundred Weight
907.18474 kg	=	1 Short Ton

OTHERS UNIT OF MEASURES

12 Articles	=	1 Dozen
12 Dozen	=	1 Gross
144 Articles	=	1 Gross
20 Articles	=	1 Score

MEASURES OF ANGLES

60 Seconds	=	1 Minute
60 Minutes	=	1 Degree
90 Degrees	=	Right Angle
180 Degrees	=	Straight Line
360 Degrees	=	1 Revolution

COOKING MEASUREMENTS

5 Milliliters	=	1 Teaspoon
10 Milliliters	=	1 Dessertspoon
15 Milliliters	=	1 Table Spoon
28.4 Milliliters	=	1 Fluid Ounce
285 Milliliters	=	1 Cup
568.26 Milliliters	=	1 Pint
1136.52 Milliliters	=	1 Quart
4546.09 Milliliters	=	1 Gallon

OVEN TEMPERATURES

Description	Celsius (C)	Fahrenheit (F)
Cool	90	200
Very slow	120	250
Slow	150 -160	300 - 325
Moderately slow	160 - 180	325 - 350
Moderate	180 -190	350 -375
Moderately hot	190 - 200	375 - 400
Hot	200 - 230	400 - 450
Very hot	230 - 260	450 - 500

THE FOUR MOST COMMON MATHEMATICAL OPERATIONS

Different ways to express each:

Multiplication → Product → Times (×)
Division → Divide → Quotient (/ or ÷)
Addition → Sum → Plus (+)
Subtraction →Minus → Take-away (-)

MATHEMATICAL SYMBOLS & SOME DEFINITIONS

+	Addition (Plus)	−	Subtraction (minus)
×	Multiplication	÷	Division
:	Is To	::	Is as
∴	Because	∴	Therefore
=	Equal To	≠	Not Equal To
≡	Identically Equal To	I-4 I	The absolute value
>	Greater Than	≥	Greater than or equal to
<	Less than	≤	Less than or equal to
√	Positive Square Root	%	Percent
{...}	Set	{ }	Empty Set or Null Set
U	Union	∩	Intersection
N	Set of Natural Numbers	W	Set of whole numbers
0.3	0.333... (repeating decimal)	(3,4)	Ordered pair
.	Decimal point	º	Degree
-5	Negative 5	+5	Positive five
10^2	Ten squared	10^3	Ten cubed
↔AB	Line AB	→AB	Ray AB
— AB	Segment AB	ABC	Plane ABC
∠	ABC Angle	M∠A	Measure of angle A
△ABC	Angle ABC	~	Is similar to
‖	Is parallel to	𝜋	Pi
Cm^2	Square centimeter	In^3	Cubic inch
3:4	Three to Four (ratio)	P(E)	Probability of Event

PROPERTIES OF ADDITION

Commutative property: When two numbers are added, the sum is the same regardless of the order in which they are added.

Example
6 + 3 = 3 + 9

Associative Property: When three or more numbers are added, the sum is the same regardless of the grouping of the numbers.

Example
(5 + 3) + 6 = 6 + (3 + 5)

Additive Identity Property: The sum of any number and zero is the original number.

Example
6 + 0 = 6

Distributive property: The sum of two numbers times a third number is equal to the sum of each number to be added times the third number.

Example
2 × (5 + 3) = 2(5) + 2(3)
The same as (2x 5) + (2x3)

PROPERTIES OF MULTIPLICATION

Commutative property: When two numbers are multiplied together, the product is the same regardless of the order of the numbers being multiplied.

Example
3 x 4 = 4 x 3

Associative Property: When three or more numbers are multiplied, the product is the same regardless of the grouping of the factors.

Example
(3 x 4) x 5 = 3 x (4 x 5)

Multiplicative Identity Property: The product of any number and one is that number.

Example
7 x 1 = 7

Distributive property: The sum of two numbers times a third number is equal to the sum of each number to be added times the third number.

Example
3 x (5 + 2) = 3 x 5 + 3 x 2

Adding Real Numbers

Positive + Positive = Positive	$6 + 3 = 9$
Negative + Negative = Negative	$(- 5) + (-2) = - 7$
Negative + Positive = Negative or Positive (***see rule***) ***Rule:*** *When adding negative and positive numbers, subtract the numbers and keep the sign of the larger number*	$(- 7) + 4 = -3$ $6 + (-13) = - 7$ $(- 6) + 8 = 2$ $5 + (-2) = 3$

Subtracting Real Numbers

Negative - Positive = Negative ***Rule:*** *A negative minus a positive is the same as same as adding two negative numbers.*	$(- 7) - 4 = -11$
Positive - Negative = Positive + Positive = Positive Note: Two negative signs next to each other makes the number to the right positive.	$4 - (-3) = 4 + 3 = 7$
Negative - Negative = Negative + Positive = Negative or Positive (***see rule***) ***Rule:*** *When subtracting a negative from a negative; neutralize the two negative signs that are together and make the number to the right positive, then subtract and keep the sign of the larger number.*	$(-7) - (-5) = (-7) + 5 = -2$ $(-5) - (-7) = (-5) + 7 = 2$

Multiplying Real Numbers

Positive x Positive = Positive	$6 \times 3 = 18$
Negative x Negative = Positive	$(-4) \times (- 2) = 8$
Positive x Negative = Negative	$5 \times (-2) = -10$

Dividing Real Numbers

Positive ÷ Positive = Positive	20 ÷ 5 = 4
Negative ÷ Negative = Positive	(-20) ÷ (-4) = 5
Negative ÷ Positive = Negative	(-20) ÷ 4 = -5
Positive ÷ Negative = Negative	20 ÷ (-4) = -5

NUMBER LINE

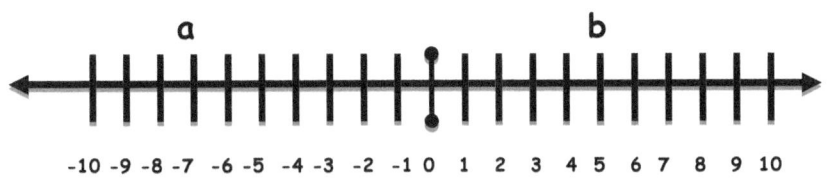

The **number line** is split into two equal halves by zero (0). Zero is neutral, so it does not get a positive or negative sign.

All numbers on the <u>line to the right of zero</u> are positive
All numbers on the <u>line to the left of zero</u> are negative
Pay special attention to how the line is numbered, in counting we go 0,1,2,3,4 etc., so it is on the number line in either direction.

Each space, be it on the right or the left counts for one, thus if we go from zero to 5 on the right it is the same amount of spaces from 0 to -5 on the left.

-7 on the number line is represented by "a" and + 5 is represented by "b" -7 + 5 = -2

This is because when you count from 0 to five on the right you have 5 spaces. From 0 to - 7 on the left side have 7 spaces. The 5 spaces on the right are cancelled out in the 7 spaces on the left, which leave 2 spaces. Because these 2 spaces are on the left side of the number line it is written as -2.

Note: the rules for addition and subtraction of real numbers will help when solving more complex problems on the number line.

VARIABLES

Variables are symbols used to represent the unknown

The **sum** of two numbers y and z: $y + z$
The **difference** of 2 numbers: $y - z$
The **product** of two numbers y and z: $(y)(z)$ or $y \times z$
The **quotient** of to numbers y and z: y/z or $y \div z$

<u>Example:</u> If **7 + y = 14** then **y** represents the unknown number that make the equation true. Therefore **7** when added to the value of "**y**" should make the left side of the equation the same amount as that on the right. This will make the equation true.

Find the value of "Y"
$7 + Y = 14$
$y = 14 - 7$
$Y = 7$

<u>Note</u>: we solve by isolating **y**, which means the **7** is moved over the equal sign. Once a number crosses the equal sign, the sign is reversed, thus the positive **7** becomes a **–7** and this is why we subtracted **7** from **14**.

To <u>prove</u> **y = 7** we substitute the **7** into the equation

    7 + y = 14; Indeed 7 + 7 = 14

Always double check!

We proved this to be correct because both sides of the equation have a value of 14.

PEMDAS - (order of solving algebraic equations)

P	Please	-	(Parenthesis) *also known as* (Brackets)
E	Excuse	-	(Exponent)
M	My	-	(Multiplication)
D	Dear	-	(Division)
A	Aunt	-	(Addition)
S	Sally	-	(Subtraction)

EXPONENTS

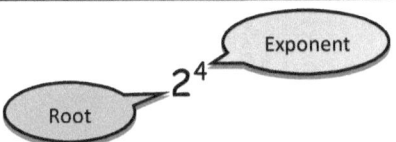

Any number raised to a power, is that number **(the root)** multiplied by itself to reflect the number of times in the exponent Eg: (2^4) = 2 x 2 x 2 x 2 = 16

Any number raised to the _zero power_ (except 0) equals 1, thus (3^0) = 1

Any number raised to the _power of one_ equals itself thus (3^1) = 3

To _multiply terms_ with the same base, add the exponents $(x^2 \, x^3)$ = x^5

To _divide terms_ with the same base, subtract the exponents $(x^6 \div x^4)$ = x^2

When a product has an exponent, each factor is raised to that power. $(xy)^a = x^a y^a$

A number with a negative exponent equals its reciprocal with a positive exponent $3\text{-}4 = \dfrac{1^4}{3}$

Laws of Exponent

$$x^a x^b = x^{a+b} \qquad (x^a)^b = x^{ab} \qquad (xy)^a = x^a y^a \qquad \dfrac{x^a}{x^b} = x^{a-b}$$

When <u>negative numbers</u> are raised to an exponent, the rule of multiplying real numbers is applied.

$(-3)^2$ = $(-3) \times (-3)$ = 9 (negative x negative = positive)

$(-3)^3$ = $(-3) \times (-3) \times (-3)$ = -18 (negative x negative = positive, that positive x negative = negative etc.)

SCIENTIFIC NOTATION

Scientific notation is an expression of the product of 10 and a number between 1 and 10. This is also a shortened way of writing very large numbers.

Example 1: 10 x 10 x 10 x 10 x 10 = 100,000 may be written as 10^5

♦ <u>Note</u> the five zeros in 100,000 represents the number (five), the exponent.

Example 2: 7.052×10^5 would be written as $7.052 \times 100,000$ = 705,200

♦ <u>Note</u> the decimal when multiplied by 100,000 was moved five times to the right, each move representing a zero (0).

SOLVING FOR AN UNKNOWN VARIABLE

When solving for an unknown variable, move all known variables to the opposite side of the equation, this will separate the unknown from the known variables.

Remember: when you move the known variable to the opposite side of the equation its sign is reversed.

Some examples:

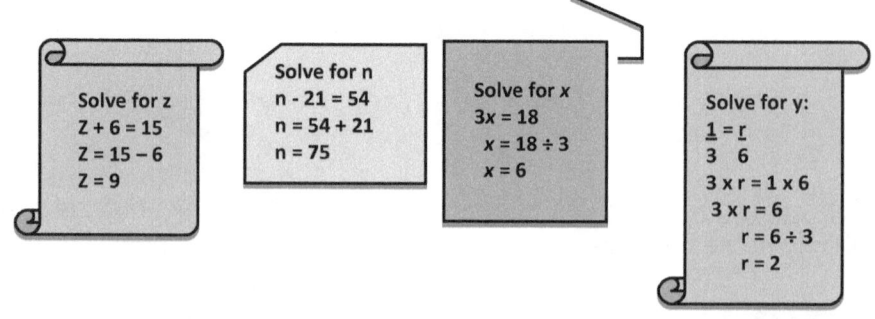

Solve for z
Z + 6 = 15
Z = 15 − 6
Z = 9

Solve for n
n − 21 = 54
n = 54 + 21
n = 75

Solve for x
3x = 18
x = 18 ÷ 3
x = 6

Solve for y:
$\frac{1}{3} = \frac{r}{6}$
3 x r = 1 x 6
3 x r = 6
r = 6 ÷ 3
r = 2

Note: in solving for "r" that we cross-multiplied.

COORDINATE PLANE

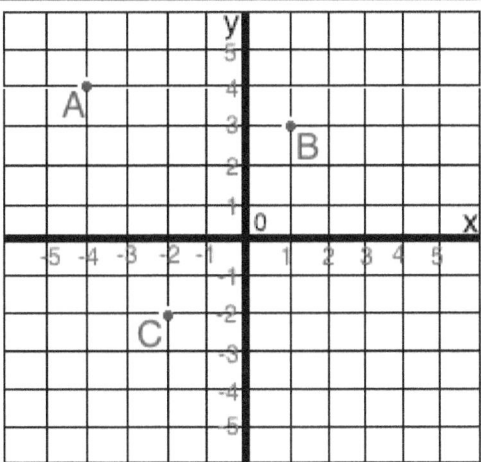

The **coordinate plane** is formed by the intersection of two number lines perpendicular to each other with the intersecting points zero, one line horizontal and the other vertical. The **intersection** at zero points is called the origin and written as (0, 0).

The **horizontal number line** is called the **x-axis** and the **vertical number line** is called the **y-axis**.

Take a look at the coordinate plane, notice **points A, B and C**. Each point is made by plotting 2 numbers on the plane that corresponds with both the **x** and **y** axis.

Point A: **-4** on the "x" axis and **4** on the "y" axis

Point B: **1** on the "x" axis and **3** on the "y" axis

Point C: **-2** on the "x" axis and **-2** on the "y" axis

Note The number on the x axis is always the first in the set, thus if asked to plot

-4 and 4 it will be written as (-4,4) so plot -4 on x and 4 on y.

SLOPE & Y-INTERCEPT

SLOPE: This term refers to the steepness of a line. The higher the line on a graph, the higher it's steepness (or slope) will be.

In order to figure out the slope of a line, you would use the equation m =<u>rise</u>
                   run

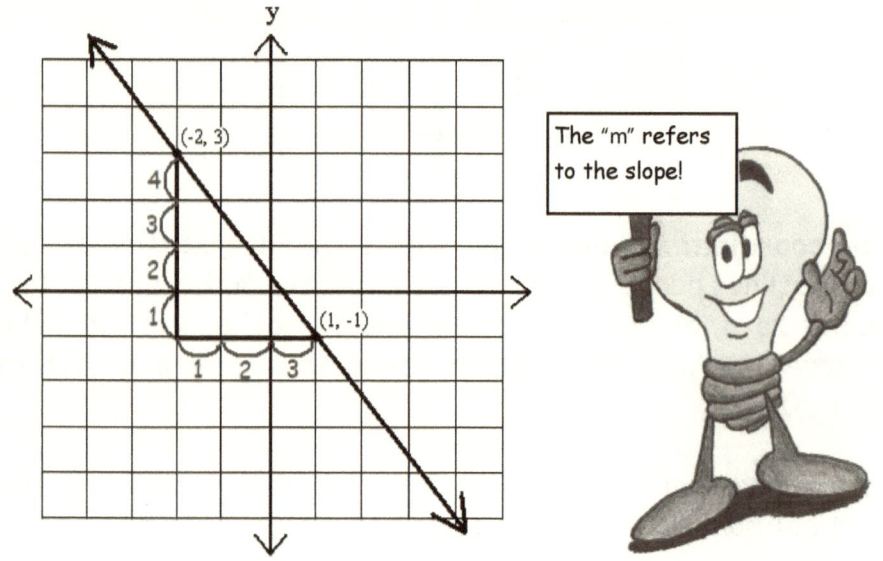

To find the slope of a line, select any point where two lines intersect to create a triangle. In this case, the rise of this triangle is the vertical line, which has four lines between two intersecting points. So the rise is four. The run is the horizontal line, which in this case has three lines between two intersecting points and therefore the run is three.

Once you have that analyzed, you can now use the formula.

m=<u>4</u> → Therefore the slope is <u>4</u>   OR 1.$\overline{3}$ if in a decimal.
    3   3

The slope of a horizontal line is always 0 and the slope of a vertical line is undefined. Two lines are parallel if they have the same slope. Parallel lines, never intersect at any point. Two lines are perpendicular if their slopes are opposite reciprocals of each other. For example, if a line has a slope of $\frac{4}{6}$,

A perpendicular line has a slope of $\frac{6}{4}$. Perpendicular lines intersect each other at right angles.

Y-INTERCEPT: The y intercept is simply where the line crosses the y-axis.

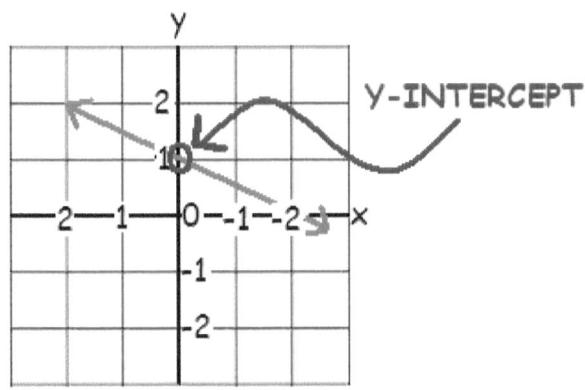

The formula used to figure out the y-intercept is <u>y = mx + b</u>. In this case, the "b" is referred to as the y-intercept.

In the graph above, the co-ordinate points are (-2, 1). Therefore the y-intercept is 1, this is because the line crossed the y axis at 1.

POSTULATES

A *line* can be extended to any length in either direction and has many points.

Through two *given points*, one and only one line can be drawn and two points determine a line.

Two lines cannot *intersect* in more than one point.

Only one *circle* can be drawn with any given point as center and the length of any given line segment as a radius.

At a *given point on a given line*, only one *perpendicular* can be drawn to the line.

For any *two distinct points*, there is only one positive real number that is the length of the line segment joining the two points.

The *shortest distance between two points* is the length of the line segment joining those two points.

A *line segment* has 2 *endpoints* and is part of a line.

An *angle* has one and only one *bisector*.

Every geometric figure is *congruent* to itself. (*Reflexive Property*)

Congruence may be expressed in either order. (**Symmetric Property**)

Two geometric figures congruent to the same geometric figure are congruent to each other. (*Transitive Property*)

Two triangles are congruent if two sides and the included angle of one triangle are congruent, respectively, to two sides and the included angle of the other. (**SAS**)

Two triangles are congruent if two angles and the included side of one triangle are congruent, respectively, to two angles and the included side of the other. (ASA)

Two triangles are congruent if the three sides of one triangle are congruent, respectively, to the three sides of the other. (**SSS**)

THEOREM

If two angles are right angles, they are congruent

If two angles are straight angles, they are congruent.

If two angles are supplements of the same angle, they are congruent

If two angles are complements of the same angle, they are congruent.

If two angles are congruent, their complements are congruent

If two angles are congruent, their supplements are congruent

If two angles form a linear pair, they are supplementary

If two lines intersect to form congruent adjacent angles, they are perpendicular

If two lines intersect, the vertical angles are congruent

EXPRESSIONS

An **expression** is a combination of numbers and variables that represents a value using mathematical operations.

<u>Examples</u> of mathematical expressions:

- ♦ $6 + 7 - \dfrac{(4 + 3)}{20}$
- ♦ $6 + 2^3$
- ♦ $m + n - p$
- ♦ $\dfrac{z - 3}{z^2 - 7}$

EQUATIONS

Equations are mathematical statements that use the equal sign to express that two sides are equal.

<u>Examples</u> of equations are as follows:

♦ $7 + 5 = 12$ $6y - 4 = 2$ $x^2 + y = -7$

INEQUALITIES

An inequality is similar to an equation in that the solution to an inequality is a value that makes the inequality true. Inequalities are solved using the same method as equations.

Note: a positive or negative number can be added to both sides of the inequality.

Any positive number can divide or multiply both sides of an inequality.

If both sides of an inequality are divided or multiplied by a negative number, take care to reverse the direction of the inequality sign.

$-4x > 24$
$-4x \div -4 > 24 \div -4$
$x < -6$
Correct answer: $x < -6$

<u>Note</u> the sign was reversed because we divided by a negative number.

CLASSIFYING POLYGONS

Triangle	3 sides	Decagon	12 sides
Quadrilateral	4 sides	Tridecagon	13 sides
Pentagon	5 sides	Tetradecagon	14 sides
Hexagon	6 sides	Pentadecagon	15 sides
Heptagon	7 sides	Hexadecagon	16 sides
Octagon	8 sides	Heptadecagon	17 sides
Nonagon	9 sides	Octadecagon	18 sides
Decagon	10 sides	Enneadecagon	19 sides
Hendecagon	11 sides	Icosagon	20 Sides

Polyhedrons are 3 dimensional shapes and the faces of the shapes are identical.

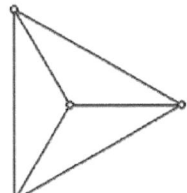   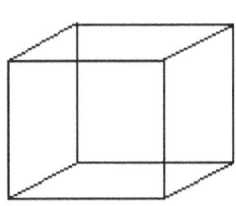   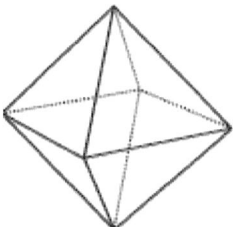

Tetrahedron (4 faces)   Cube (6 faces)   Octohedron (8 faces)

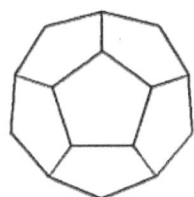   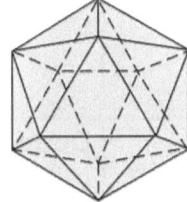

Decahedron (12 faces)   Icosahedron (20 faces)

GEOMETRICAL FIGURE

Shape	Name	Angles and Sides
	Parallel Lines	No angles, no sides
	Line	Made up of many points and extends continuously in either direction.
	Ray	Has one endpoint and extends continuously in one direction.
	Line segment	Part of a line and has 2 endpoints
	Acute Angle	Measure less than 90 degrees
	Right Angle	Measure 90 degrees
	Obtuse Angle	Measure more than 90 degrees
	Right Triangle	Has 3 sides and 3 angles, 1/4 turn triangle measures 90^0

	Square	Has 4 right angles and 4 equal sides, measures 360^0
	Rectangle	Has 4 right angles, parallel sides are congruent (same) and measures 360^0
	Circle	Has no angles or sides, measures 360 degrees
	Oval	Has no angles or sides
	Circular Ring	Has no angles
	Parallelogram	Has opposite sides with the same length and they are parallel.
	Trapezium	Has 4 angles and one pair of parallel sides.
	Pentagon	Has 5 angles and five sides

	Hexagon	Has 6 angles and 6 sides
	Octagon	Has 8 angles and 8 sides
	Sphere	Has no angles
	Cone	Has one edge
	Cylinder	Has two identical flat circular ends and the side is curved.
	Cube	Made up of 6 squares and has 12 edges

	Rectangular Prism	6 sides and all angles are right angles.	
	Rhombus	Has 4 equal sides with equal length. Opposite sides are parallel and the angles are equal.	

Symmetry and Transformations

Reflection

A reflection is a way of transforming a shape as a regular mirror does at home. In a plane, the result of reflecting an object in a mirror line (or an axis of reflection) is called its mirror image. The objects and image are symmetrical about the mirror line.

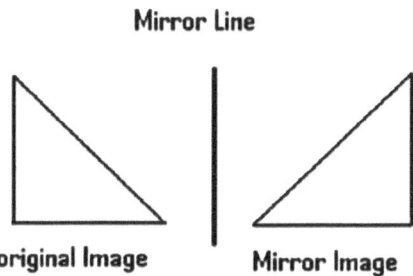

Mirror Line

original Image   Mirror Image

Note: The *reflexive property of equality* just says that a = a: anything is congruent to itself: the equals sign is like a mirror, and the image it "reflects" is the same as the original.

Symmetry

Symmetry is a type of pattern that a shape has. It deals with the exact matching of a position or form about a point, line or plane.

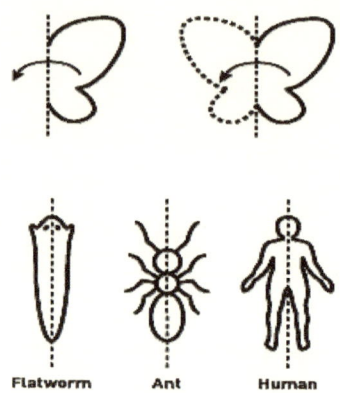

Flatworm　　　Ant　　　Human

Note: The *symmetric property of equality* says that if a = b, then b = a.

Translation

A movement along a straight line is called a translation. A plane is said to have a translational symmetry if it can be translated and still looks the same.

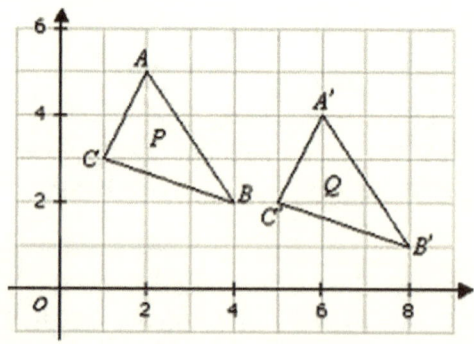

Note: the *transitive property of equality* says that if a = b and b = c, then a = c

TYPES OF TRIANGLE

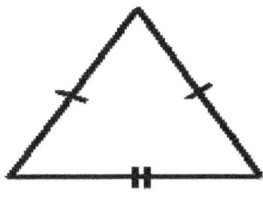

Isosceles (Have 2 sides equal)

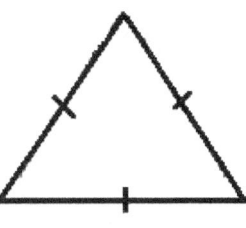

Equilateral (All sides are equal)

Scalene (No sides are equal)

Pythagoras Theorem: $A^2 + B^2 = C^2$ Example: $3cm^2 + 4cm^2 = h$ (c)

$3cm^2 + 4cm^2 = 7cm^2$

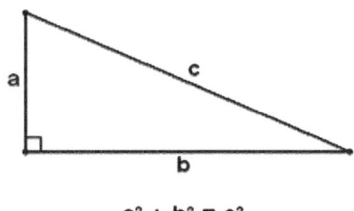

$a^2 + b^2 = c^2$

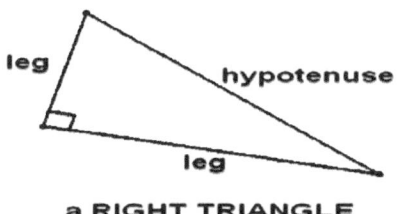

a RIGHT TRIANGLE

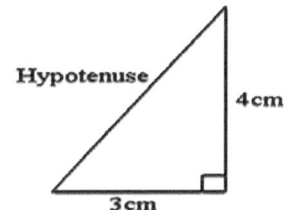

GEOMETRIC CALCULATIONS

Find the third Angle of a Triangle

The sum of the <u>interior angles</u> of a triangle are equal to 180^0

To find the third angle of a triangle when the other two angles are known, Subtract the sum of the two known angles from 180^0

Example: How many degrees are in the third angle of a triangle whose other two angles are 40^0 and 65^0? **Answer:** $180^0 - (40^0 + 65^0) = 75^0$

Find the Fourth Angle of a Quadrilateral

The sum of the interior angles of a quadrilateral are equal to 360^0. To find the fourth angle of a quadrilateral when the other three angles are known, subtract the sum of the degrees in the three angles from 360^0.

Example: How many degrees are in the fourth angle of a quadrilateral whose other three angles are 80^0 and 110^0 and 95^0? **Answer:** $360^0 - (80^0 + 110^0 + 95^0) = 75^0$

Supplementary Angles

Two angles are supplementary if the sum if their angles equal 180^0. If one angle is known, its supplementary angle can be found by subtracting the measure of its angle from 180^0.

Example: What is the supplement of 143^0? **Answer:** $180^0 - 143^0 = 37^0$

Complementary Angles

Two angles are complementary if the sum of their angles equal 90^0. If one angle is known, its complementary angle can be found by subtracting the measure of the angle from 90^0.

Example: What is the complementary angle of 43^0? **Answer:** $90^0 - 43^0 = 47^0$

FORMULAS

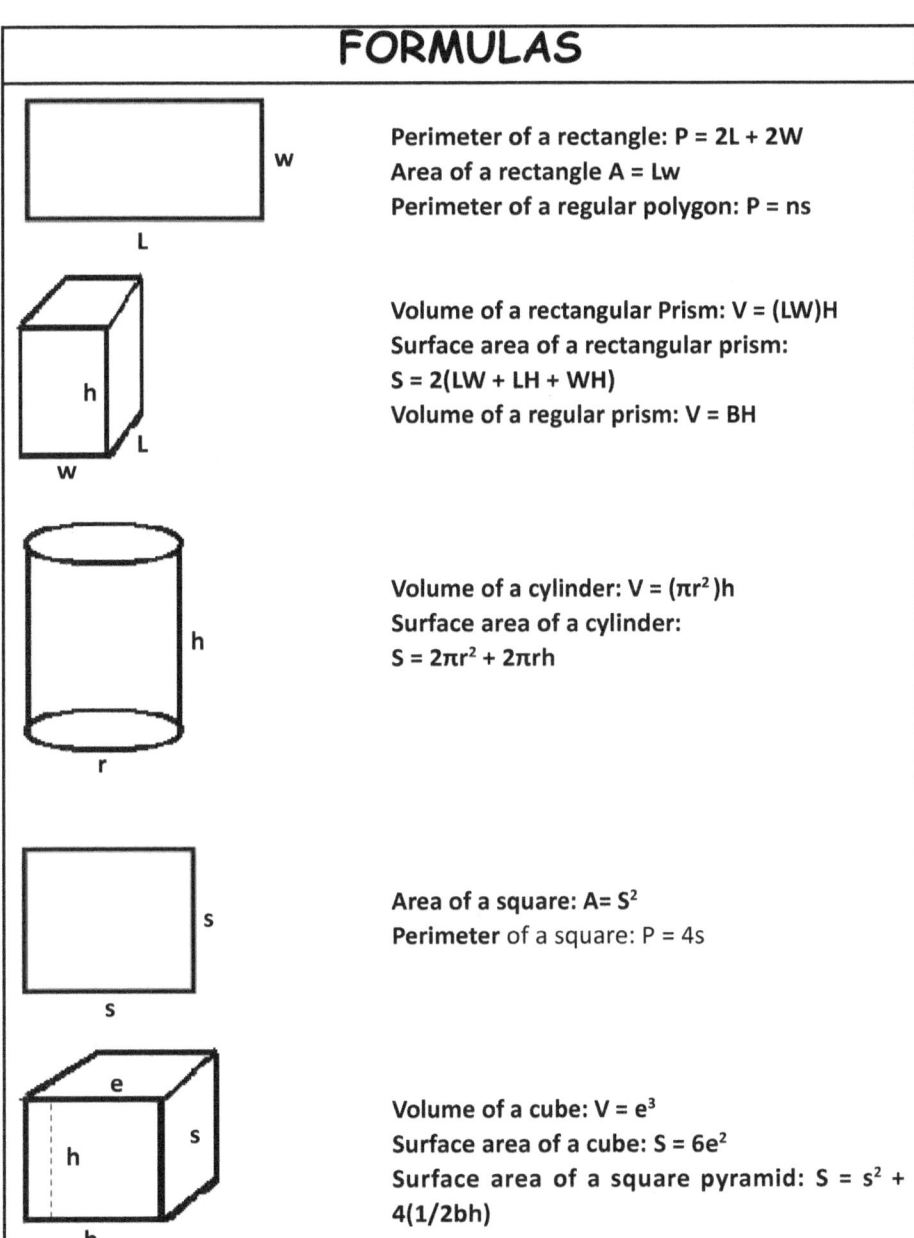

Perimeter of a rectangle: P = 2L + 2W
Area of a rectangle A = Lw
Perimeter of a regular polygon: P = ns

Volume of a rectangular Prism: V = (LW)H
Surface area of a rectangular prism:
S = 2(LW + LH + WH)
Volume of a regular prism: V = BH

Volume of a cylinder: V = $(\pi r^2)h$
Surface area of a cylinder:
S = $2\pi r^2 + 2\pi rh$

Area of a square: A= S^2
Perimeter of a square: P = 4s

Volume of a cube: V = e^3
Surface area of a cube: S = $6e^2$
Surface area of a square pyramid: S = s^2 + 4(1/2bh)

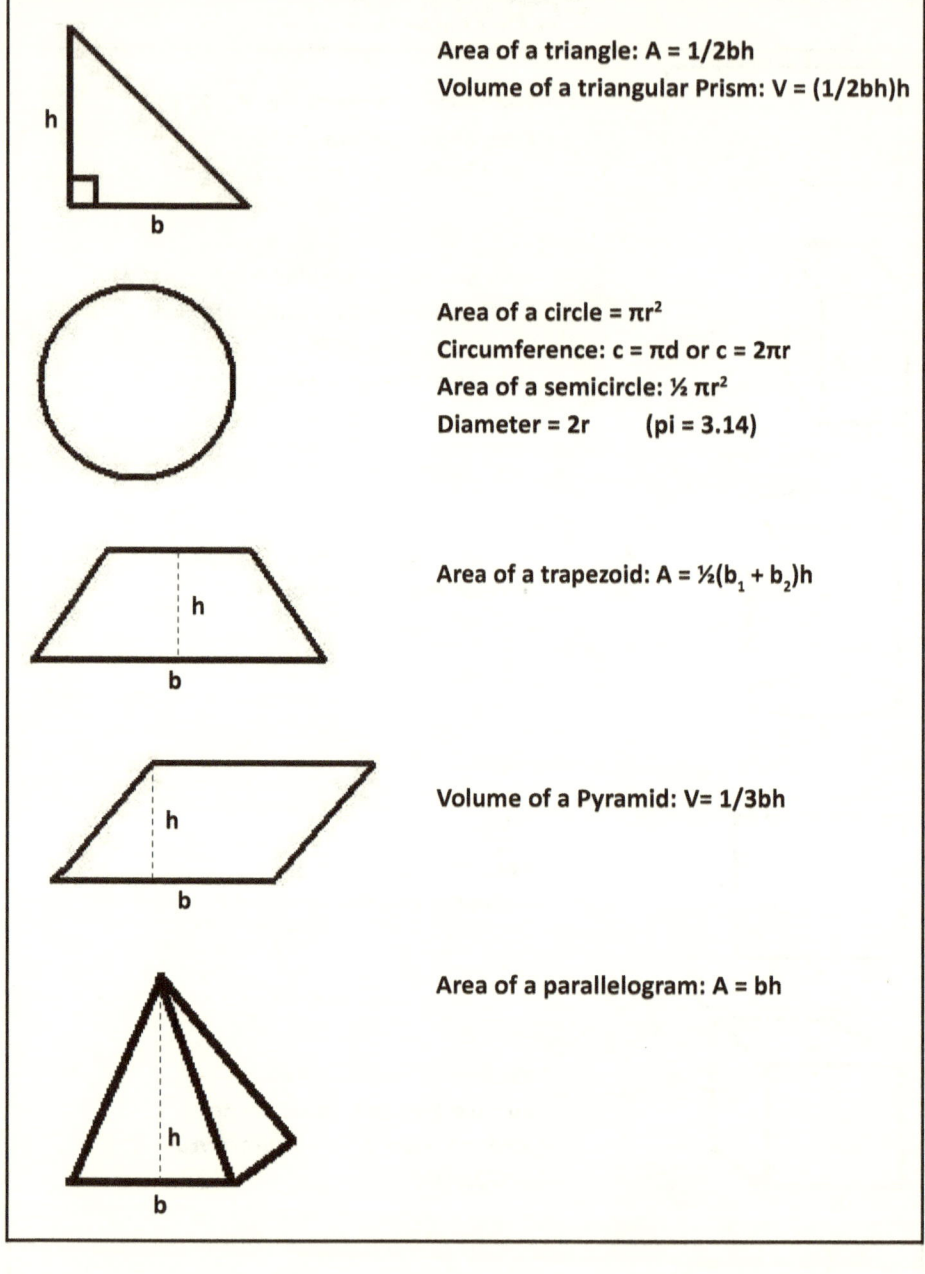

Area of a triangle: A = 1/2bh
Volume of a triangular Prism: V = (1/2bh)h

Area of a circle = πr²
Circumference: c = πd or c = 2πr
Area of a semicircle: ½ πr²
Diameter = 2r   (pi = 3.14)

Area of a trapezoid: A = ½(b₁ + b₂)h

Volume of a Pyramid: V= 1/3bh

Area of a parallelogram: A = bh

OTHER FORMULAS

Temperature conversion formula

$$F = \frac{9}{5}C + 32.$$

To convert degrees **Fahrenheit to Celsius**, subtract 32 and multiply by 5/9

To convert degrees **Celsius to Fahrenheit**, multiply by 5/9 and add 32

Celsius (^{0}C) C = 5/9(F-32)

Fahrenheit (^{0}F) F = 9/5 C+32

Pythagoras Theorem: $A^2 + B^2 = C^2$

Distance = Speed (rate) x Time (d=rt or d = st)

Average Speed (rate) = <u>distance traveled</u>
time taken

Time = distance ÷ Speed (rate) (t = d/s)

Simple interest = Principal x rate x time (I=prt)

Discount = list price x rate of discount (d =lp x r of d)

Sale price = regular price - discount (sp= rp-d)

Sales tax = marked price x rate of sales tax (t = mp x r of t)

Total cost = marked price + sales tax (tc = mp + t)

Commission = Total Sales x rate of commission (c = ts x r of c)

Miscellaneous Notes:

Pure water freezes at 32 °F = 0 °C
Pure water boils at 212 °F = 100 °C
Body temperature 98.6 °F = 37 °C

STATISTICS & PROBABILITY

PART 1: STATISTICS

<u>Statistics</u> is a branch of mathematics in which the collection and interpretation of numbers are compared to see patterns in the results. Your attendance record at school is an example of statistics. This helps your teachers on your participation in your classroom.

RANGE, MEDIAN, MODE and *MEAN* are the basic vocabulary of statistics.

RANGE is the difference between the greatest and least number in a set of data.

For example, look at this climate graph for **Vancouver, British Columbia, Canada.**

VANCOUVER	JAN	FEB	MAR	APR	MAY	JUN	JUL	AUG	SEP	OCT	NOV	DEC
Temperature (°C)	3	5	6	9	12	15	17	17	14	10	6	4
Precipitation (mm)	150	124	109	75	62	46	36	38	64	115	170	179

What is the temperature range in Celsius?
To figure out the answer you must find the lowest number and subtract that from the highest number.
3°C ←Lowest temperature

17°C ←Highest temperature

<u>SOLVE:</u> *17-3=14°C*

→ *Therefore the range between the temperatures each month is 14°C. This means that Vancouver has a temperature range of 14°C.*

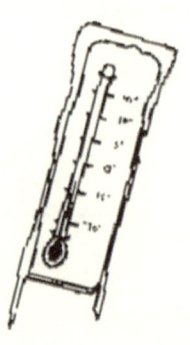

MEDIAN is the middle number in a set of data. To find the median, place the numbers in order from least to greatest. The number in the middle would be the median.

Let's take a look at the climate graph for Vancouver again:

VANCOUVER	JAN	FEB	MAR	APR	MAY	JUN	JUL	AUG	SEP	OCT	NOV	DEC
Temperature (°C)	3	5	6	9	12	15	17	17	14	10	6	4
Precipitation (mm)	150	124	109	75	62	46	36	38	64	115	170	179

In order to find what the mean in this set of data is, you must arrange all the numbers from least to greatest, and then find the number that is in the middle.

E.g.   3, 4, 5, 6, 6, (9), 10, 12, 15, 17, 17

These are all the numbers for the temperature throughout the year in Vancouver, B.C in order.

3, 4, 5, 6, 6   9, 10   12, 15, 17, 17

There are 5 numbers on either side of the nine, making it the middle number (or the median).

→ *Therefore, the 9°C is the median temperature.*

If the set of data was even, the two numbers that are in the middle of the rest of the data is the median. If you went further into your search to find the median, you add both numbers together and divide them by 2, since there are two numbers.

In an even-numbered set of data, the median is sometimes recognized as the two numbers in the middle of the set.

E.g. If the data was: **3, 4, 5, 6, 6, (9, 10,) 12, 15, 17, 17, 20**
Solve→ *9+10=19*
          19÷2=9.5
→ *Therefore the median is 9.5°C*

MODE is the number that appears most frequently in a set of data. Some set of data may not have a mode; others may have two or more modes.

Let us take a look at the climate graph for Vancouver and try to identify any modes in the temperature:

VANCOUVER	JAN	FEB	MAR	APR	MAY	JUN	JUL	AUG	SEP	OCT	NOV	DEC
Temperature (°C)	3	5	6	9	12	15	17	17	14	10	6	4
Precipitation (mm)	150	124	109	75	62	46	36	38	64	115	170	179

On the graph we have circled the mode(s) in the temperature. You may notice that there are two modes, 6 and 17. This is figured out by identifying which number(s) appear more than once. In this case it is the numbers 6 and 17.

→ *Therefore the mode(s) are 6 and 17*

MEAN is the average number in a set of data.

Take a look at the climate graph for Vancouver once more:

VANCOUVER	JAN	FEB	MAR	APR	MAY	JUN	JUL	AUG	SEP	OCT	NOV	DEC
Temperature (°C)	3	5	6	9	12	15	17	17	14	10	6	4
Precipitation (mm)	150	124	109	75	62	46	36	38	64	115	170	179

In order to find the mean (or the average) number in a set of data, add all the numbers, then divide them by the amount of numbers there are.

3 + 5 + 6 + 9 + 12 + 15 + 17+ 17 + 14 + 10 + 6 + 4 ←
In total, there are 12 numbers
$\quad\quad$ **= 118**
$\quad\quad$ Next you take the sum and divide by the amount of numbers there are. In this case, it is 12.
$\quad\quad$ **118÷12= 9.8**
$\quad\quad$ → *Therefore the mean number (or temperature) is 9.8 (°C)*

PART 2: PROBABILITY

Probability means the chance or likelihood that something will happen. In math, probability is the numerical chance that a specific outcome will occur. That number is always between zero and one. **Zero (0)** means that there is *NO* chance of that event occurring. **One (1)** means that the event is certain to occur. The closer the probability is to one, the greater the chance of that event occurring.

Probability can be expressed in two ways: in *ratios* or in *percentages*.

For example, if you tossed a coin, you would have a one in two chance of tossing heads → **1:2** *OR* **50%**

It is a one in two chance because there are only two sides.

OR

If you had a paper bag filled with four rubber balls: 1 red ball, 2 green balls, and 1 yellow ball. The probability of you pulling out a green ball would be two in four →**2:4** *OR* **50%**

GLOSSARY

Acute Angle - measures less than 90°

Adjacent angles - two angles in the same plane that have a common vertex and common side but do not have interior points in common

Average - one number or value that represents a whole group

Area - the amount of space inside the boundary of a flat (2-dimensional) object

Angle Bisector - line segment that bisects (cut) one of the vertex angles of a triangle or a line in two halves

Chord - A line segment linking any two points on a circle

Circle - A line that forms a closed loop; every point on which is a fixed distance from a center point

Center - the point in the very center of the circle

Circumference - the distance around the circle

Commission - a fee for services paid to a sales person based on a percentage of the amount of money made from a sale

Coordinate Plane - a plane formed by the intersection of a horizontal and a vertical number line. The horizontal number line is called the *x*-axis and the vertical number line is called the *y*-axis. The number lines intersect at their zero points. This point of intersection is called the origin and written as (0, 0).

Complementary Angles - two angles whose sum measures 90° degrees

Composite numbers - has other factors other than itself and one

Cone - A solid three-dimensional object that has a circular base and one vertex

Cylinder - A solid object with two identical flat ends that are circular, has one curved side with the same cross-section from one end to the other

Congruent - two things having the same shape and size

Cube - a three-dimensional shape with six square or rectangular sides

Decimal - Is a proper fraction whose denominator is a power of 10

Denominator - Is the bottom number in a fraction

Diameter - The distance across the circle also known, as the length of any chord passing through the center and it is twice the radius

Distance - Describes how far apart two objects are

Even Numbers - these are numbers exactly divisible by two

Expanded Form - the way to write a number that shows the sum of values of each digit of a number

Expression - a mathematical statement that represents numerical values

Exponent - a number that shows how many times the root number should be multiplied

Equation - a mathematical statement that two expressions are equal

Equilateral triangle - A triangle with all three sides of equal length

Factors - Factors are the numbers multiply together to get another number

Fraction - is a number that represents part of a whole

Greatest Common Factor (GCF) - The highest number that divides exactly into two or more numbers

Hexagon - **A six-sided polygon**

Inequality - An inequality says that two values are not equal

Intersect - The point where two lines meet or crosses

Improper Fraction - has the numerator larger than the denominator

Interior angles - The area between the rays that make up an angle, and extending away from the vertex to infinity

Isosceles triangle - A triangle with two of its sides equal in length

Line - is made up of many points and extend in either direction continuously

Line Segment - a part of a line with two end points.

Linear pair of angles - two adjacent angles whose sum is a straight angle

Lowest Common Multiple (LCM) - The smallest number that is a multiple of two or more numbers

Mid-point- the middle of a line (halfway)

Mixed Fraction - consist of a whole number and a proper fraction

Multiple - The result of multiplying by an integer (a positive or negative whole number or zero)

Natural Numbers - these are counting numbers

Numerator - is the top number in a fraction

Obtuse Angle - an angle that is greater than 90° but less than 180°

Octagon - eight-sided polygon

Odd Numbers - are not exactly divisible by two

Proper Fraction - has the numerator smaller than the denominator

Proportion - is formed by two equal ratios

Parallel lines - Two lines on a plane that never meet and are always the same distance apart

Parallelogram - A quadrilateral with both pairs of opposite sides parallel

Percent - is proportion in relation to a whole (usually the amount per hundred)

Perpendicular - Lines that are at right angles (90°) to each other (parallel lines)

Pentagon - a five-sided polygon

Perimeter - is the distance around the edge if a shape

Pi - constant, approximately equal to 3.142 used in determining the area of a circle

Place Value - is the value of where the digit is in the number, such as the units (ones), tens, hundreds, etc.

Prime numbers - are divisible only by themselves and one except for the number one

Profit - The positive gain from an investment after subtracting for all expenses

Radicals - An expression that has a square root, cube root, etc. (the symbol is $\sqrt{}$)

Reciprocal - is a fraction turned upside down (the numerator switches place with the denominator)

Radius - the distance from the center to any point on the circle

Rate - is a ratio that expresses how long it takes to do something (*speed*)

Ratio - is a comparison of two numbers

Ray - A portion of a line that starts at a point and goes off in a particular direction to infinity

Rectangle - A 4-sided polygon where all interior angles are 90°

Rectangular prism - A solid (3-dimensional) object that has six faces that are rectangles.

Reflection - to create a reflection, a figure is flipped to create a mirror image.

Rhombus - A four-sided shape where all sides have equal length

Right angle - an angle that is equal to 90°, one quarter of a full revolution

Right triangle - A triangle that has a right angle (90°)

Rotation - the figure is rotated around a fixed point

Sales price - The *price* of a good or service that is being offered at a discount

Sales tax - tax based on the cost of the item purchased and collected directly from the buyer

Scalene triangle - A triangle with all sides of different lengths

Simple interest - Interest paid only on the original principal

Slope - describes its steepness, incline, or grade of a line

Sphere - A 3-dimensional object shaped like a ball

Square - A 4 sided polygon with 4 right angles and 4 sides of equal length

Squared Number - The number you get when you multiply another number by itself

Square Root - a number that when multiplied by itself equals a given number

Supplementary Angles - two angles, the sum of whose degrees measures 180°

Symmetry - one shape that looks exactly like another if flipped, or turned

Translation - the movement of an object along a straight line without turning

Trapezium - a quadrilateral with no parallel sides

Variable - A symbol for an unknown number represented by a letter like x or y.

Vertical angles - are two angles in which the side of one angle is opposite ray to the side of the second angle

Volume - is the space occupied by the three-dimensional shapes and is expressed by cubic units

Marsha Smith is a private math tutor of students in Middle School. She normally tutors boys and girls in grades 6, 7 and 8. As a result, she realizes the great demand for this quick reference study guide and has been prompted by both parents and students to compile this manual.

As a tutor with over 15 years experience, she has assisted numerous students who have been struggling with mathematics, to find a clearing at the end of the maze. Working with students from different age groups and from a variety of backgrounds has allowed her to utilize different, effective teaching strategies, in order to turn the mathematical lights on in these young minds.

Her hope is that once this mathematical flame has been lit, it will remain burning for years to come, so the children will continually want to have mathematics as part of their daily lives.

Her principle is based on understanding the core challenge of each student, then working with them to find a solution. She has experienced many successes in this regard and has even received many compliments from thankful parents and teachers. To her, there is no greater joy, than seeing each child's eyes light up with understanding and the beautiful smile that usually fill their faces from ear to ear when they know they have got it right.

So, for the student, here is a manual that summarizes the necessary mathematical concepts that you need for grades 6,7,8 and beyond. Parents, I know your child will benefit immensely from this resource, as many have done in the past. So if you are reading this manual, I want to say, "Thank you!"